The Tools of Scientists

by Ellen Ungaro

Table of Contents

What tools do scientists use?
20
105
6
34

A Scientific Adventure at Sea

The scientists on the research ship BRP *Hydrographer Presbitero* were crowded around a video monitor. They were watching images taken at the bottom of the ocean. Suddenly they spotted an unusual-looking sea creature. It was bright orange, about 10 centimeters (3.94 inches) long, and had ten long tentacles. At first, the scientists thought it might be a squid or a shrimp. It wasn't. The scientists realized it was a new species. One never seen before! The scientists called it the "squidworm."

This discovery was just one of many from a scientific research expedition to the Celebes Sea in the Pacific Ocean. The scientists spent ten days exploring the sea. Their goal was to learn more about sea life and identify new species. How did they do it?

They used a variety of tools. They used scuba equipment to observe the marine animals in their natural habitat. They carried jars to capture specimens, or samples, of marine life. Twice a day, giant nets were used to scoop up plankton. Plankton are the tiny living things that form the bottom of the food chain in the ocean.

▲ These nets, called bongo nets, are made of very fine mesh. Only tiny plankton are trapped in the net; larger fish are not usually captured.

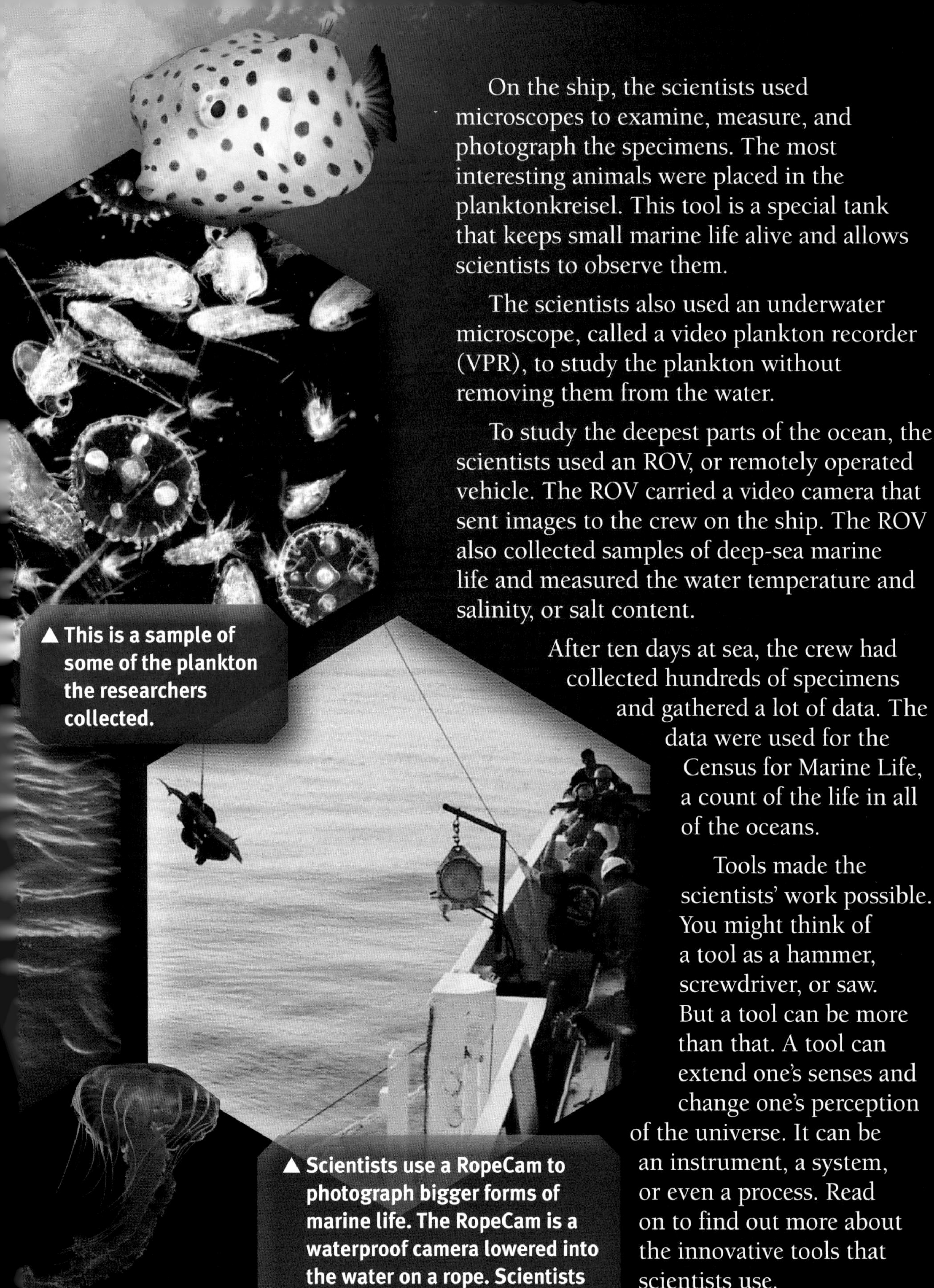

On the ship, the scientists used microscopes to examine, measure, and photograph the specimens. The most interesting animals were placed in the planktonkreisel. This tool is a special tank that keeps small marine life alive and allows scientists to observe them.

The scientists also used an underwater microscope, called a video plankton recorder (VPR), to study the plankton without removing them from the water.

To study the deepest parts of the ocean, the scientists used an ROV, or remotely operated vehicle. The ROV carried a video camera that sent images to the crew on the ship. The ROV also collected samples of deep-sea marine life and measured the water temperature and salinity, or salt content.

After ten days at sea, the crew had collected hundreds of specimens and gathered a lot of data. The data were used for the Census for Marine Life, a count of the life in all of the oceans.

Tools made the scientists' work possible. You might think of a tool as a hammer, screwdriver, or saw. But a tool can be more than that. A tool can extend one's senses and change one's perception of the universe. It can be an instrument, a system, or even a process. Read on to find out more about the innovative tools that scientists use.

▲ This is a sample of some of the plankton the researchers collected.

▲ Scientists use a RopeCam to photograph bigger forms of marine life. The RopeCam is a waterproof camera lowered into the water on a rope. Scientists tie bait to the camera to attract sea life.

Chapter 1

Tools for Measuring

What tools are used for measuring in science?

It's Monday morning. After hitting the snooze button a few times, you finally roll out of bed to get ready for school.

You wash your face and brush your teeth. You check the thermometer to see how cold it is outside, and then you get dressed.

You pour a bowl of cereal, add just the right amount of milk, and cut up some fruit for breakfast.

After you eat, you put your notebooks, textbooks, and some pens and pencils into your backpack and head out the door.

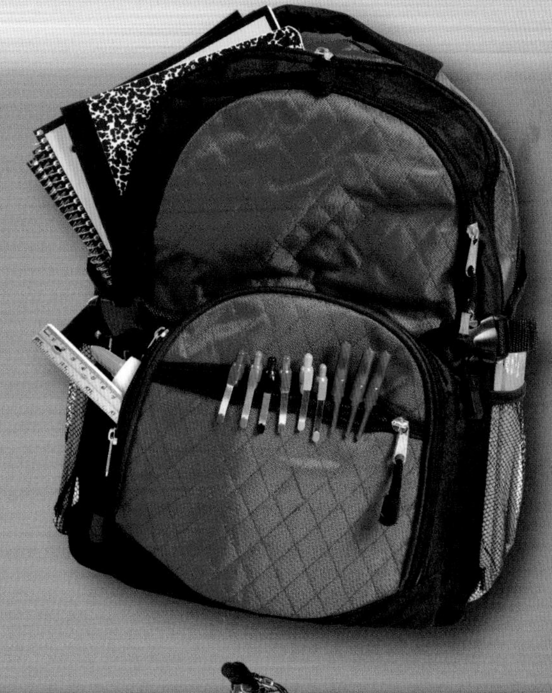

Essential Vocabulary

After putting on your helmet, you jump on your bike and head for school.

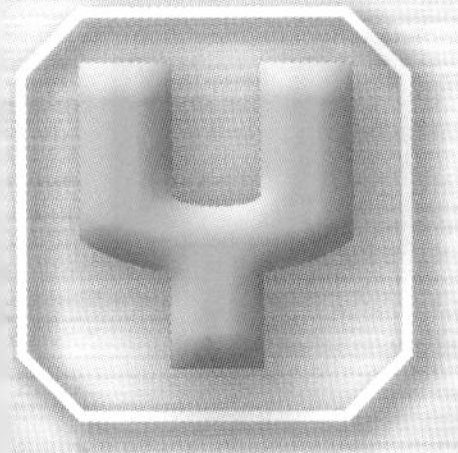

You are using tools in all of these actions. You use tools to help you get ready for school—the alarm clock, toothbrush, thermometer, and utensils you needed to eat. You use tools to safely get to school—your helmet and your bike. You use tools at school—notebooks, textbooks, pens, and pencils. A tool can be an instrument, device, system, or process used to accomplish a task. There are many tools people use each day.

Some of the most common tools are used for measuring. These tools measure quantities such as length, mass, weight, temperature, and time. Carpenters use rulers to measure the dimensions of boards for building. Cooks use measuring cups and spoons to make sure they add the right volumes of ingredients. Chemists use balances to measure the exact masses of substances in a chemical reaction.

The Metric System

As you just read, a tool does not need to be an instrument or device. A tool can also be a system. The **metric system** is one of the most important tools of scientists. Developed in 1790 by the French Academy of Science, the metric system was the first standardized system of measurement.

Before that time, people had been using all sorts of different measurements, many of them based on body parts. An inch was the width of a thumb. A foot was the length of a foot. A yard was the distance from the tip of a nose to the fingertips of an outstretched hand. As you might guess, these measurements varied greatly from one person to another. This made talking about measurements confusing, and buying and selling goods difficult.

People with different occupations also developed different types of measurement. Farmers measured length in furlongs, a unit based on the length of a furrowed row of land. Sailors measured the depth of water in fathoms. Different trades used "barrel" as a measurement.

With the development of the metric system, these varied and confusing measurements were eliminated. In their place was a simple and easy-to-use system based on the number ten and multiples of ten. The metric system also uses a set of prefixes that can be added to a unit to indicate its relative size.

Common Metric Prefixes

Prefix	Meaning
micro-	1/1,000,000 or 0.000001
milli-	1/1,000 or 0.001
centi-	1/100 or 0.01
deci-	1/10 or 0.1
kilo-	1,000
mega-	1,000,000

The table above shows common metric prefixes and their meanings. Because the metric system is based on the number ten, it is easy to convert from one unit to another by multiplying or dividing by ten.

In 1960, a revised version of the metric system known as the International System of Units (SI) was adopted. Today, most countries throughout the world use this system. One notable exception is the United States, which uses the customary system. However, scientists everywhere use the SI. The SI has base units for length, mass, temperature, and time. All other SI measurements can be derived from the base units.

Quantity	SI Base Unit	Symbol
length	meter	m
mass	kilogram	kg
temperature	kelvin	K
time	second	s

Why do you think it is important that all scientists use the same SI units of measurement?

Length

The SI base unit of length is the **meter** (m). The average height of a doorknob from the floor is one meter.

A meter is divided into smaller units called decimeters, centimeters, and millimeters. A decimeter (dm) is one-tenth (1/10 or 0.1) of a meter. There are 10 decimeters in a meter. The diameter of an orange is about 1 dm.

A centimeter (cm) is one-hundredth (1/100 or 0.01) of a meter, so there are 100 centimeters in a meter. The width of a shirt button is about 1 cm.

Metric Conversion Table for Length

Metric		Customary	
10 mm	1 cm	0. 39 in	0.0328 ft
100 cm	1 m	3.28 ft	1.0936 yd
1000 m	1 km	1093.6 yd	0.62137 mi

A millimeter (mm) is one-thousandth (1/1,000 or 0.001) of a meter. There are 1,000 millimeters in a meter. A dime has a thickness of about 1 mm.

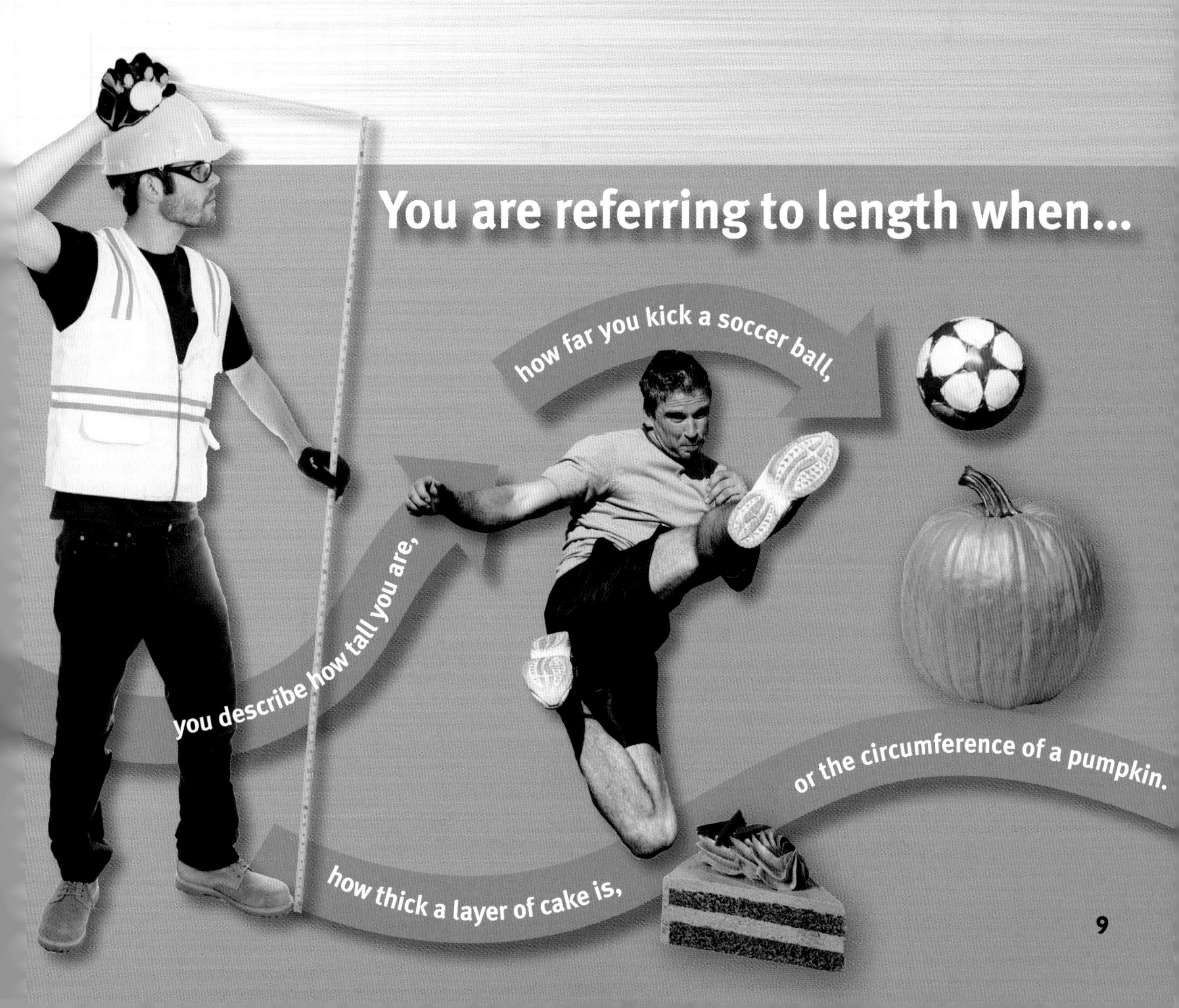

Measurements of longer lengths are usually made in kilometers (km). One kilometer is equal to 1,000 meters. One kilometer is almost two-thirds of a mile. Mount Everest, the tallest mountain in the world, has a height of almost 9 km. The flying distance from New York City to Chicago is approximately 1,150 km.

Length can be measured with a metric ruler or a meterstick.

Mass

All forms of matter have **mass**. Mass is a measure of the quantity of matter. The SI base unit of mass is the **kilogram** (kg). The kilogram is the only base unit in SI that has a prefix. A guinea pig has a mass of about 1 kg.

How to Use a Metric Ruler

1. First identify the markings on the ruler. Most metric rulers have both centimeter and millimeter marks. The longer lines are the centimeter marks and the shorter lines are the millimeter marks. A millimeter is one-tenth (1/10) of a centimeter.
2. Place the ruler on a flat surface. Carefully line up one end of the object with the first mark on the ruler.
3. Find the line closest to the other end of the object.
4. Count how many whole centimeters the object is. In the photo above, the end of the key falls between the 4-centimeter and 5-centimeter mark.
5. Then count the number of millimeters after the 4-centimeter mark. The end of the key is even with the third millimeter mark. The key is 4.3 centimeters. How many millimeters is the key?

Other commonly used units of mass are the gram and milligram. One gram (g) is one-thousandth (1/1,000) of a kilogram. A dollar bill has a mass of about 1 g. There are 1,000 grams in a kilogram. A milligram (mg) is one-thousandth (1/1,000) of a gram, or one-millionth (1/1,000,000) of a kilogram. Ten grains of salt have a mass of 1 mg.

Metric		Customary	
1,000 mg	1 g	0.03527 oz	0.00022 lb
1,000 g	1 kg	35.274 oz	2.2046 lb

Mass can be measured by comparing the object with standard masses. A triple beam balance is one instrument for measuring mass.

How to Use a Triple Beam Balance

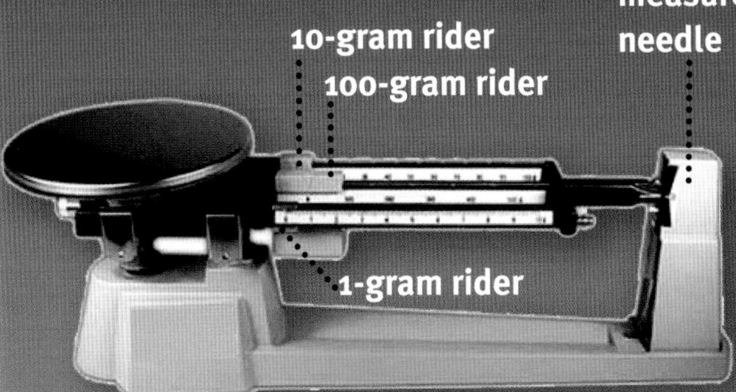

1. Calibrate the balance. First set all three riders to zero. Then adjust the calibration knob under the pan until the measurement needle aligns exactly to zero.

2. Obtain the "tare weight." If you are using a container to hold your object, use the steps below to weigh the empty container first. Record this "tare weight." Then repeat the following steps with the target object in the container.

3. Place the container on the center of the pan. Move the rider on the 100-gram beam to the right until the pointer falls below the balance mark. Then move it back one notch on the beam.

4. Move the rider on the 10-gram beam to the right until the pointer falls below the balance mark. Then move it back one notch on the beam.

5. Slowly move the rider on the 1-gram beam until the pointer is even with the balance mark.

6. Record the readings on the three beams, then add them together. If you used a container, subtract the tare weight to determine the mass of the object.

How to Use a Spring Scale

1. Identify the units of measurement on the spring scale. Many spring scales include both grams and newtons.

2. Make sure the pointer is at zero. Place the object on the hook. Wait for the object to settle.

3. Read the gram or newton marking next to the pointer to find the weight.

newtons

grams

Weight

In everyday language, mass and weight are used interchangeably. However, mass and weight are not the same thing. Weight is a measure of the pull of gravity on an object. Weight can change depending on location. Because the pull of gravity on the moon is one-sixth the pull of gravity on Earth, a person will weigh one-sixth of his or her Earth weight on the moon. An object's mass does not change unless matter is added to or removed from the object. The person will have the same mass on Earth as on the moon.

The metric unit of weight is the **newton** (N). The newton is a unit of force, which is what the pull of gravity is. Because the pull of gravity on Earth does not vary much, kilograms are also used as a unit of weight. Weight is measured with a spring scale. Bathroom scales and scales at supermarkets are usually spring scales.

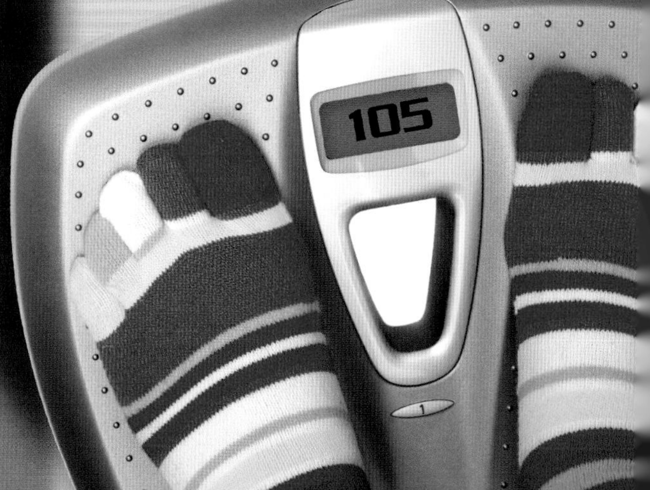

Volume

Volume is the amount of space an object takes up. The tool used for measuring volume and the units used for the measurement depend on the sample being measured. The three basic types of samples are liquids, rectangular solids, and irregular solids.

Liquid

The unit of volume for a liquid is the **liter** (L). Although the liter is not an SI unit, it is accepted for use with the SI because it is important and widely used. The other units of liquid volume are the deciliter (dL), centiliter (cL), and milliliter (mL). A liter is equal to either 10 deciliters, 100 centiliters, or 1,000 milliliters. A small juice box would have a volume of about 1 dL. Two teaspoons of liquid equal 1 cL. A milliliter is about one-fifth of a teaspoon. A kiloliter (kL) is used to measure larger volumes. An Olympic-size swimming pool contains 2,500 kL of water. The most commonly used measurements for volume are the milliliter and the liter.

The volume of a liquid can be measured using a graduated cylinder. Graduated cylinders come in different sizes, from 10 mL to 1,000 mL.

Metric		Customary	
	1 mL	0.03381 fl oz	0.264 gal
10 mL	1 cL	0.3381 oz	0.0423 cups
10 cL	1 dL	3.381 oz	0.423 cups
10 dL	1 L	33.814 oz	4.227 cups
	1 L	1.0567 qts	0.2642 gal

How to Use a Graduated Cylinder

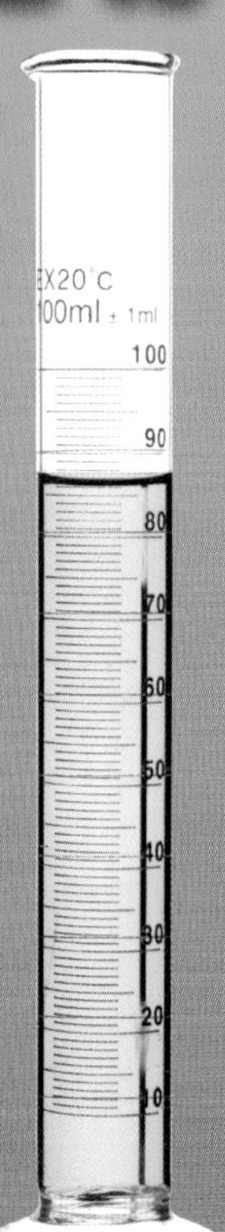

1. Start by identifying the markings on the graduated cylinder. A graduated cylinder is usually calibrated in milliliters. This means that each line on the graduated cylinder is one milliliter.

2. Rest the graduated cylinder on a flat surface. Then pour the liquid into the graduated cylinder.

3. Notice that the surface of the liquid is slightly curved. The curve is called the **meniscus** (meh-NIS-kus). To determine the volume of the liquid, you must read the millimeter marking at the bottom of the meniscus.

How to Find the Volume of a Rectangular Solid

1. Use a metric ruler to measure the dimensions (length, width, and height) of the block in centimeters.
2. Record the value of each measurement in centimeters.
3. Calculate the volume. Multiply the three dimensions using the formula $v = l \times w \times h$.
4. Include the units in your answer. You are multiplying units as well as values, so the answer is in cubic centimeters (cm^3). For example, 2 cm x 2 cm x 3 cm = 12 cm^3.

3 cm

2 cm

2 cm

▲ The volume of this block is 12 cm^3. What would the volume be in milliliters?

Rectangular Solid

The volume of a rectangular solid is determined by measuring the length, width, and height of the solid with a metric ruler, and then multiplying the three values. This calculation can be written as the formula:

volume = length x width x height

$$v = l \times w \times h$$

Because each dimension is measured in centimeters, the unit of volume is **cubic centimeters** (cc or cm^3). The average sugar cube has a volume of 1 cubic centimeter. The average shoe box is 4,320 cubic centimeters.

In the SI, 1 cubic centimeter is equal to 1 milliliter. This makes it easy to convert volume measurements from cubic centimeters to milliliters.

Irregular Solid

Not all solids have a rectangular shape. Think of objects such as stones and shells. The length, width, and height of such objects cannot be measured with a metric ruler. How, then, can the volume of an object that is neither a liquid nor a rectangular solid be measured? For nonporous objects, the method of water displacement can be used. A nonporous object cannot be permeated by liquid. A sponge is porous. A coin is nonporous.

Science Tools: Measuring Lung Health

One way for a doctor to determine the health of a person's lungs is to measure the volume of air the person can hold in his or her lungs. Doctors use a tool called a spirometer to make this measurement. After taking a deep inhalation, the person exhales into the spirometer and the doctor measures the volume of air exhaled in one second. This shows how well the lungs' passageways are functioning.

How to Use the Water Displacement Method

1. Select the smallest possible graduated cylinder that can hold your object.
2. Fill a graduated cylinder about half full with water. Record the volume of water in the cylinder by reading the marking that corresponds to the bottom of the meniscus.
3. Carefully place the irregular solid into the cylinder. Make sure the solid is completely covered by water. Record the volume of the solid and the water combined.
4. Subtract the volume of the water from the combined volume of the solid and water. This number is the volume of the irregular solid. The unit for this volume is cubic centimeters, because that is the unit used to measure the volume of solids.

Finding the Right Tool for the Job

Time Required

45 minutes

Group Size

two or three students

Materials

- 5 graduated cylinders ranging in size (10 mL, 25 mL, 50 mL, 100 mL, 250 mL)
- 25-mL beaker

Procedure

1. Fill the beaker with water.
2. Transfer 5 mL of water from the beaker to a 250-mL graduated cylinder.
3. Then pour the 5 mL of water in the 250-mL graduated cylinder into a 10-mL graduated cylinder.
4. Carefully read and record the amount of water seen in the 10-mL graduated cylinder.
5. Repeat steps 1–3 above but begin with the 100-mL graduated cylinder.
6. Repeat steps 1–3 above but begin with the 50-mL graduated cylinder.
7. Repeat steps 1–3 above but begin with the 25-mL graduated cylinder.
8. Compare the 4 final measurements recorded in step 4.

Analysis

1. Describe how you would measure 15 mL of water needed for an experiment.
2. What general statement (rule) might you make about the laboratory equipment used in an experiment?
3. Why would you not use a 4-gallon pot to make spaghetti for just you and your sister?

Here's a trick question for you: Which has a greater mass, a kilogram of bricks or a kilogram of feathers? Did you fall for the trick and answer "bricks," as most people do? A kilogram of bricks has the same mass as a kilogram of feathers. What if the question were: Which has a greater volume, a kilogram of bricks or a kilogram of feathers? The answer would be feathers because you would need a lot more feathers than bricks to reach the same mass. This is because the amount of mass in the volume of a single feather is less than the amount of mass in the volume of a single brick.

The amount of mass in a given volume is called **density**. Density is defined as the mass per unit volume of a substance. The formula for density is:

$$\text{Density} = \frac{\text{mass}}{\text{volume}} \quad \text{or} \quad d = \frac{m}{v}$$

The units of density are grams per milliliter or grams per cubic centimeter. Density is an important property of matter. The density of water is 1 g/mL. Objects with a density less than that of water will float on water. Objects with a density greater than water will sink in water.

▼ Ever wonder why most woods float in water? These wood blocks all have different densities. The blocks with a greater density than water sink. The blocks with a lesser density float.

▲ The density of the oil in this test tube is less than the density of the water.

Temperature

Temperature is a measure of the average kinetic energy with which the particles in an object move. Temperature can be measured on different scales. The **Celsius** scale is commonly used in science. On this scale, temperature is measured in degrees Celsius (°C). Water freezes at 0°C and boils at 100°C. These temperatures are known as the freezing point and boiling point.

In the SI, the base unit of temperature is the **kelvin** (K). On this scale, the degree sign is not used. Kelvin units are the same size as degrees Celsius. However, each value is 273 degrees greater than the Celsius degree. The freezing point of water on the Kelvin scale is 273 K. The boiling point is 373 K.

The temperature scale you are probably familiar with is the Fahrenheit scale. On this scale, the freezing point of water is 32°F and the boiling point is 212°F.

◀ **This thermometer measures temperature in both Celsius and Fahrenheit degrees.**

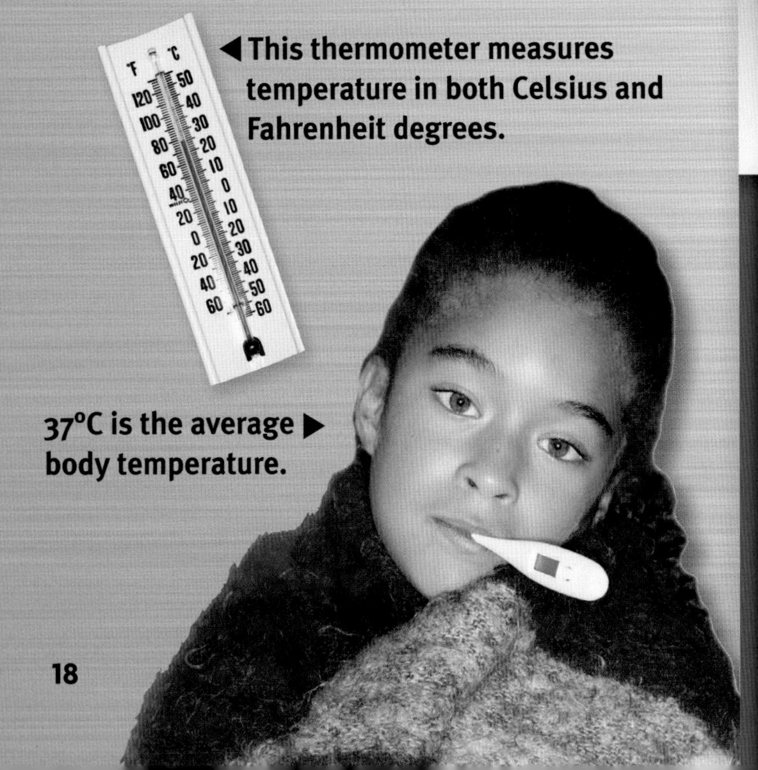

37°C is the average ▶ body temperature.

Time

The base unit of time in all measurement systems, including SI, is the second (s). Although hours and days are not SI units, they are accepted units of time. A kilosecond is 1,000 seconds. A millisecond is one-thousandth (1/1,000) of a second. Instruments such as clocks, watches, and stopwatches are used to measure time.

▼ **A stopwatch can be used to measure time.**

They Made a Difference: Lord Kelvin

The Kelvin temperature scale was named for the Irish-born physicist and mathematician William Thomson, also known as Lord Kelvin, who helped define absolute zero. Absolute zero is the temperature at which all particle motion theoretically stops. Absolute zero, 0 K, is equivalent to –273°C. Conversions: $C = 5/9(F-32)$, $F = (9/5)C + 32$, and $K = C + 273$

SUMMING UP

- Since ancient times, people have developed systems and tools to help them measure things.
- The SI system was developed so that there would be a standard, easy-to-use system of measurement with base units for length, mass, temperature, and time.
- SI base units include the meter, kilogram, and the second. The Kelvin scale and the Celsius scale are the units of measure for temperature. The newton is the unit of measure for weight.
- We can use a meterstick or metric ruler to measure length, a triple beam balance to measure mass, a thermometer to measure temperature, and a stopwatch to measure time.
- The volume of liquids can be determined with a graduated cylinder. The volume of a rectangular solid can be derived using the formula $v = l \times w \times h$.
- Water displacement is used to find the volume of a nonporous irregular object.
- Density, which is the amount of mass in a given volume, can be derived by the formula $d = m/v$.

Putting It All Together

Choose one of the activities below. Work independently, with a partner, or in a small group. Share your presentation and what you have learned with the class.

1. The United States is one of the few countries that still uses the customary measurement system. Research the history of the SI system in the United States and the efforts to convert the country to metric. Write a paragraph reporting your findings.

2. Work with a partner to create a visual summary of metric measurement. Include objects illustrating all the measurements you have read about and their units. For example, you might show a bottle of juice that has a volume of 1 liter and a nickel that has a mass of 5 grams.

3. Work with a small group to create a five-question quiz that tests what tools and SI units you would use to measure different objects.

EUREKA!

CARTOONIST'S NOTEBOOK
ILLUSTRATED BY JOEL CARLSON

TODAY IN LAB, WE'LL BE USING GRADUATED CYLINDERS TO MEASURE NONPOROUS IRREGULAR SOLIDS...

EUREKA!

I WAS JUST REFERRING TO ONE OF MY FAVORITE BEDTIME STORIES ABOUT ARCHIMEDES, THE GREEK MATHEMATICIAN AND MY PERSONAL HERO, FATHER OF THE WATER DISPLACEMENT THEORY. THE STORY BEGINS IN ANCIENT GREECE...

HERE WE GO!

DENSITY IS MASS DIVIDED BY VOLUME. ARCHIMEDES COULD EASILY WEIGH THE CROWN TO FIND THE MASS, BUT HE NEEDED THE VOLUME IN ORDER TO FIND DENSITY. HE COULD NOT DAMAGE THE CROWN OR MELT IT DOWN INTO A REGULAR SHAPE SO HE COULD MEASURE ITS DENSITY. ARCHIMEDES WAS STUMPED.

WHAT ARE SOME OTHER WAYS THAT ARCHIMEDES COULD HAVE SOLVED THIS PROBLEM? WHAT ARE SOME OTHER TOOLS HE MIGHT HAVE USED? YOUR MIND IS AN IMPORTANT TOOL. WHY DO YOU THINK THE MIND IS THE GREATEST SCIENTIFIC TOOL OF ALL?

Chapter 2

Tools for Observing

What tools do scientists use to observe and identify objects?

The SI system, triple beam balance, spring scale, graduated cylinder, and thermometer are all tools scientists use to measure. There is more to science than measurement, however. Scientists also observe, describe, and identify things in the world around them. And, as you might guess, they have tools for all of these jobs. Some tools are as simple as hand lenses that can be tucked into a scientist's pocket. Others are complex tools that help scientists search the ocean floor or examine rock samples on distant planets.

Telescopes

Have you ever looked up at the night sky on a clear night? If so, you were probably amazed at all there was to see—billions and billions of stars. A careful observer would be able to pick out the planets Mars and Venus among all the twinkling stars.

Essential Vocabulary

binoculars	page 26
hand lens	page 27
microscope	page 27
reflecting telescope	page 24
refracting telescope	page 24
sonar	page 31
telescope	page 24

These objects are millions of kilometers away. To get a clearer view, scientists use **telescopes**. Most telescopes work by collecting light. They are able to gather and focus much more light than the human eye can, making it possible to see objects that are faint and far away. There are two basic types of telescopes—**refracting telescopes** and **reflecting telescopes**.

A refracting telescope uses lenses to gather light. This type of telescope has two clear lenses—a convex objective lens at the end of the telescope and a second magnifying lens in the eyepiece. One problem with refracting telescopes is that the images can have a fuzzy, rainbow halo around the edges.

A reflecting telescope uses mirrors to gather light. It has a concave objective mirror at one end and a smaller mirror near the eyepiece. Reflecting telescopes do not distort images in the same way that refracting telescopes do. That is why most telescopes used today are reflecting telescopes.

Today many of the telescopes scientists use are computerized so that they can be precisely focused on a location in space. Many have cameras that record images. A scientist no longer has to spend hours looking through a telescope lens. Instead, she can review and analyze images at any time. In addition, collecting information this way lets scientists in different places "use" the telescope.

The Hubble Space Telescope, which is in orbit around Earth, is one of the most powerful telescopes working today. Because the Hubble orbits Earth beyond Earth's atmosphere, it can see farther into space. Images from the Hubble are beamed down to Earth.

Through early telescopes, scientists were able to study the surfaces of the moon and Mars, find the rings around Saturn, and see the moons around Jupiter.

How a Telescope Works

Refracting Telescope

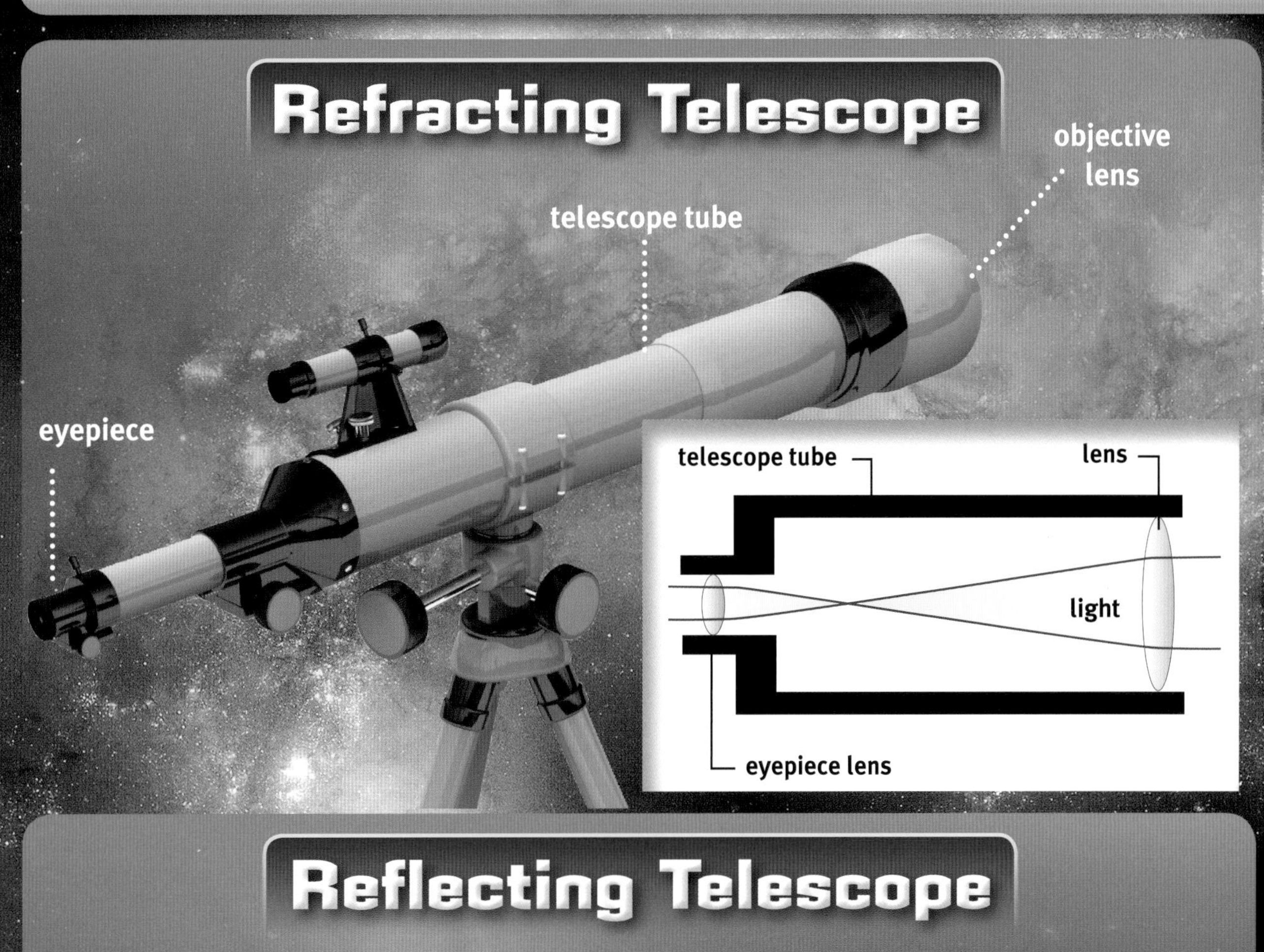

Reflecting Telescope

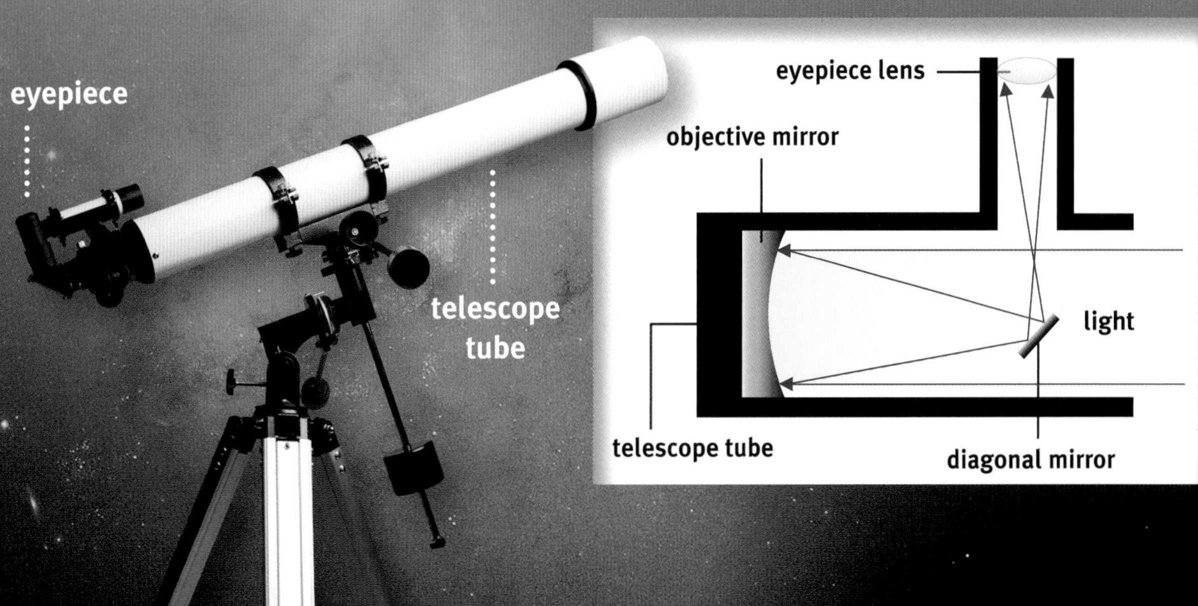

The First Telescope

In 1608, Hans Lippershey, a Dutch eyeglass maker, created what many believe was the first telescope when he put two lenses together in a long tube. The tool made distant objects seem larger and closer. At first, it was called the spyglass because it was perfect for spying on one's enemies. However, in 1609, the scientist Galileo Galilei had another idea. After tweaking the design, Galileo used it to study the night sky. He discovered that Earth's moon has craters and mountains. He discovered the four moons of Jupiter. He also saw that Venus has phases just like Earth's moon. The scientist Isaac Newton later improved Galileo's telescope, replacing one of the lenses with a group of mirrors. Newton's telescope was a reflecting telescope.

With the help ▶ of his telescope, Galileo was able to find evidence that Earth revolved around the sun.

▲ Binoculars help scientists observe animals in their natural habitats without disturbing them.

Binoculars

Binoculars are another tool scientists can use to observe objects that are far away. Binoculars work on the same principle as a telescope. In fact, binoculars are really two telescopes side by side. Like a refracting telescope, binoculars have objective lenses that gather and focus light.

Checkpoint: Read More About It

See more amazing images and find out more about the Hubble Telescope at the NASA Web site.

▼ This image taken from the Hubble shows two galaxies colliding.

The lens in each eyepiece produces a magnified image of an object. In general, you cannot see as far with binoculars as you can with telescopes. Binoculars do, however, have other advantages. You can see through both eyes and you see a wider field of vision, or area. So while a telescope can zoom in on one crater on the moon, binoculars allow you to see more of the area around the crater.

Binoculars are also portable, or easy to carry. Many biologists working in the field use them to observe animals in their natural habitats. A researcher studying the rain forest uses binoculars to observe animals living in the treetops. At sea, whale watchers use binoculars to focus on the whales' tails. The tail is one way to identify an individual whale. Backyard astronomers also use binoculars.

Hand Lenses and Microscopes

Scientists also need to see small objects up close so that they can observe the details. A geologist in the field might examine an unknown rock. She might want to note its color. She might want to see if it is dull or shiny. What tool would help the geologist identify the other characteristics of the rock? She would use a **hand lens**. The lens will magnify the rock. Using the lens, the geologist can see the mineral makeup of the rock and correctly identify it. Biologists also use a hand lens in the field to examine everything from a plant specimen to an unusual insect.

Microscopes are not as portable as hand lenses are, but they offer much more powerful magnification. Just like telescopes, there are different types of microscopes.

Light or optical microscopes may be nothing more than a single lens. These are called simple microscopes. However, a single lens can magnify only so much. This limitation led to the invention of the compound microscope. Like the refracting telescope, the compound microscope uses two lenses, the objective lens and the eyepiece. The total magnification of a compound microscope is equal to the product of the magnifications of each of the two lenses.

▼ **A hand lens can help scientists identify lichens and other organisms when working in the field.**

For example, if an objective lens can magnify 40x and an eyepiece can magnify 10x, then the total magnification will be 40x multiplied by 10x, which equals 400x. The maximum magnification of a light microscope is about 1,000x. However, there are many objects, such as viruses and cell structures, that require a magnification much greater than 1,000x. To solve this problem, scientists invented electron microscopes, which use beams of electrons instead of light. Just like your television at home, the beams of electrons produce an image on a screen.

▼ **Compare these images of red blood cells. The image on the right was taken through a light microscope and the image on the left was taken with a scanning electron microscope (SEM).**

The Fathers of Microscopy

Robert Hooke (1635–1703) and Anton van Leeuwenhoek (1632–1723) are two important figures in the world of microscopes. Hooke, an English scientist, designed and built microscopes. In 1665, he published *Micrographia*, a book of illustrated descriptions of objects he had seen with his microscope. These included insects, bird feathers, and cork. It is from his descriptions of cork that the term *cell* was coined to describe the features of plant tissues.

At the same time in Holland, Anton van Leeuwenhoek was also building microscopes. Leeuwenhoek's microscopes had only one lens—they were really just powerful magnifying glasses. But Leeuwenhoek was very good at making lenses, so his images were clearer than Hooke's. Leeuwenhoek examined anything he could with his microscope. He even studied scrapings from people's teeth, making him the first person to describe bacteria. He was also the first to describe protozoans, animal-like microscopic organisms, which he observed in a sample of pond water. Robert Hooke later confirmed Leeuwenhoek's discoveries.

How a Compound Microscope Works

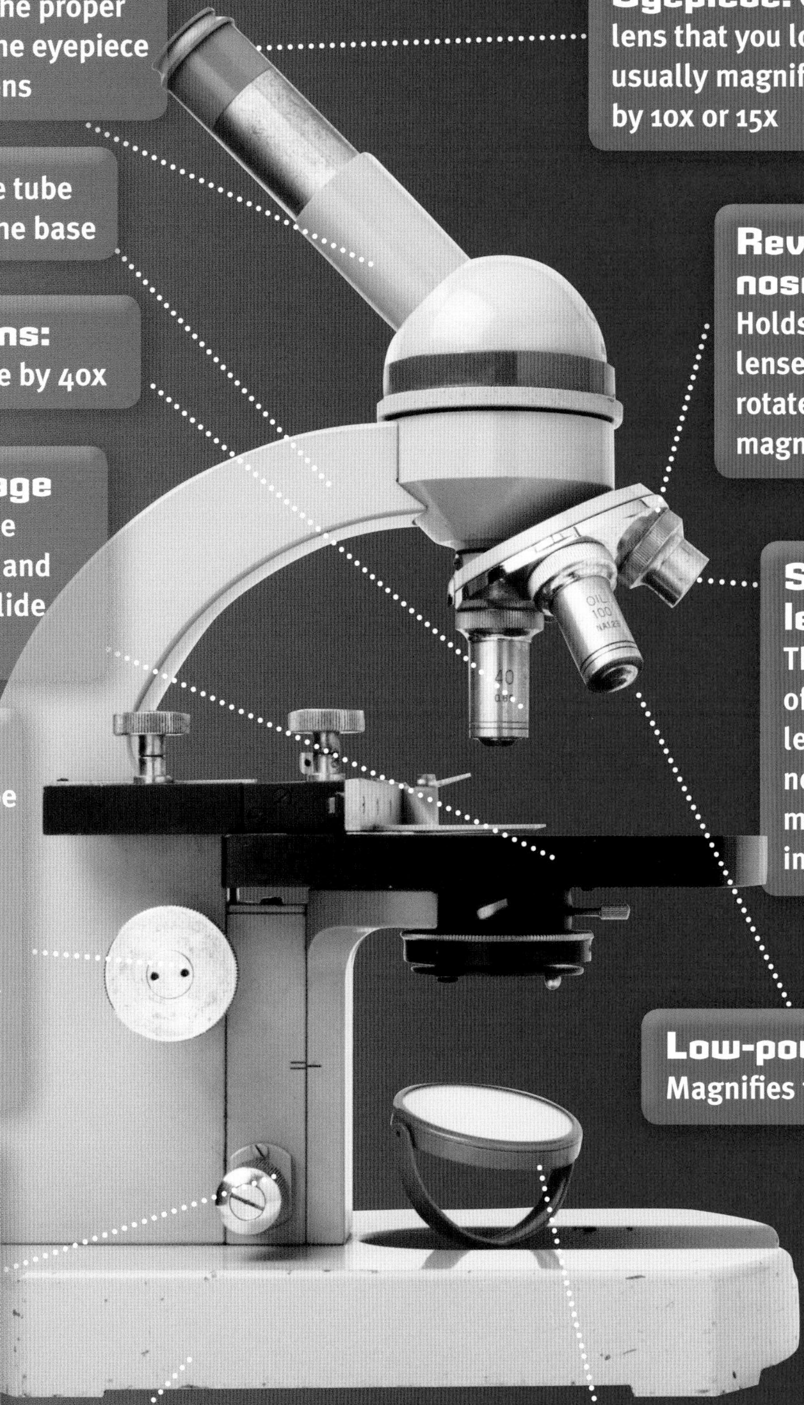

Tube: Maintains the proper distance between the eyepiece and the objective lens

Arm: Supports the tube and connects it to the base

High-power lens: Magnifies the image by 40x

Stage with stage clips: Supports the slide being viewed, and the clips keep the slide in position

Coarse focus: Moves the body tube up and down and should be the first adjustment knob used (never move down while looking through the microscope)

Fine focus: Moves the body tube slightly and is the knob used to sharpen the focus

Base: Supports the microscope

Eyepiece: Contains the lens that you look through; it usually magnifies the image by 10x or 15x

Revolving nosepiece: Holds the objective lenses and can be rotated to change magnification

Scanning lens: The smallest of the objective lenses on the nosepiece, it magnifies the image 4x

Low-power lens: Magnifies the image 10x

Light source: A steady source of illumination

Pond Water Investigation

Time Required

30 minutes

Group Size

two or three students

Materials

compound microscope, lens paper, pond water (protozoa culture), slides, coverslips, toothpicks, eyedroppers, protozoa reference guide

Procedure

1. Obtain a small jar of pond water (protozoa culture) from your teacher.
2. Wipe a glass slide with a piece of lens paper and place it on the desk in front of you.
3. Using an eyedropper, transfer a single drop of the pond water (protozoa culture) to the center of the slide.
4. Place a coverslip over the culture using the following technique:
   a) Holding the edges of the coverslip between your thumb and forefinger, place the edge of the coverslip in the drop. This will cause the water to spread.
   b) Support the top edge of the coverslip with a toothpick and slowly lower it until it is as close to the surface of the slide as possible.
   c) Remove the toothpick and allow the coverslip to fall the remaining distance.
5. Place the slide on the stage, turn the rotating nosepiece so that the 10x (low power) objective lens is in place above the coverslip, and scan the slide for moving organisms.
6. When you find an area with several organisms, turn the nosepiece so that the 40x (high power) objective is in place. (Never use your coarse adjustment when you are on high power.)
7. Sketch as many different organisms as possible and refer to the reference guide to identify those organisms.
8. Label each organism and any structures you can identify under high power.

Analysis

1. How many different types of organisms did you see? What colors did you observe? What shapes did you observe?
2. What characteristics or structures did you observe that seemed to help the protozoa move?

Sonar

Sonar, or *so*und *na*vigation *a*nd *r*anging, is a tool that uses sound to detect, locate, and measure distances of underwater objects. Sonar sends a sound signal into the water. Sound, in the air or in the water, travels as waves. When the sound waves hit an object in the water, the signal is reflected, or bounced back. By measuring the time between when the signal is sent out and when the reflected sound is received, the distance to the object can be calculated. For example, suppose that four seconds go by between the sending out of a signal and the return of its echo. This means that it takes the sound two seconds to travel to the submerged object and two seconds to return. The average speed of sound in water is about 1,500 meters per second. So if it takes two seconds for the sound to reach the object, the object is 3,000 meters away.

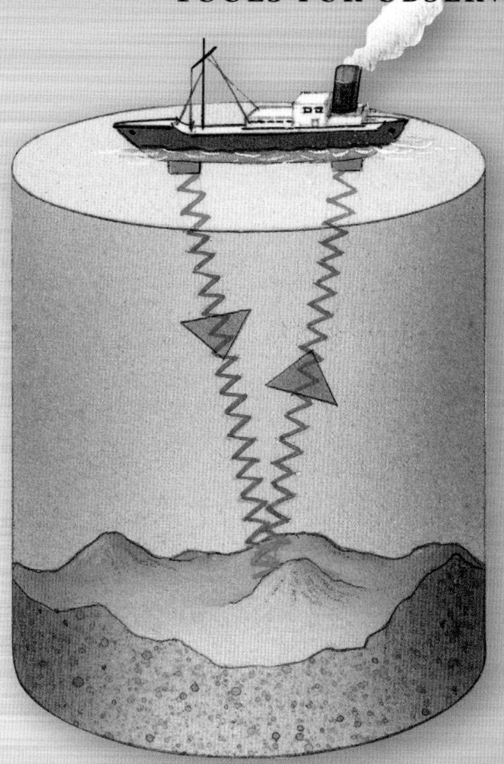

▲ **Ships can use sonar to determine the depth of water they are traveling in.**

▲ **Sonar helps with navigation and can also be used to track the direction of ocean currents.**

Sonar has a variety of applications. Scientists use sonar to map the depth of water. They also use sonar to locate and identify objects under water. When a sonar signal bounces back after hitting an object, the quality of the sound changes. An experienced sonar technician can tell whether a sonar echo is being produced by the ocean floor, a school of fish, a rock outcrop, or a whale.

The speed of sound also changes with changes in the temperature of water. Sound travels more slowly through cold water than warm water. So scientists can use sonar to measure water temperature. Sonar can also be used to track the direction of ocean currents and record seismic events, such as small earthquakes, under water.

Rovers

How do scientists gather data about planets or plant and animal life deep in the ocean? They use remotely operated vehicles (ROVs). These tools allow scientists to make observations and collect information about places that humans cannot visit.

The Mars Exploration Rovers, named Spirit and Opportunity, have been exploring the planet since 2004. The Rovers are packed with tools. They carry cameras that take pictures of the sky and the surface of the planet. The Rovers also have a Microscopic Imager, which is both a microscope and a camera. The Rock Abrasion Tool drills into Martian rocks. The spectrometers help scientists determine the mineral makeup of the rocks.

Remotely operated vehicles are also used to study deep ocean water. One ROV named Jason has been used to study everything from the hydrothermal vents on the ocean floor to ancient shipwrecks. Scientists on research vessels lower Jason to the ocean floor. The cameras that Jason carries send images up to the scientists working on the surface. Jason is also able to record temperatures and collect samples of mud, water, rocks, and marine animals. Jason can collect up to 136 kilograms (300 pounds) of rock samples at a time.

It takes three people at the remote controls to operate Jason when it is in the water. ▶

▼ Scientists at NASA direct the movement of the Mars Exploration Rovers from Earth.

SUMMING UP

- Scientists have tools to help them observe, describe, and identify things.
- Telescopes help scientists study objects in space, while binoculars help scientists observe objects on Earth.
- A hand lens is a portable tool that scientists can use when they are working in the field.
- Scientists use light microscopes and electron microscopes to see objects that aren't visible with the naked eye.
- Sonar uses sound to locate objects and measure distance.
- ROVs explore places that scientists cannot visit.

Putting It All Together

Choose one of the activities below. Work independently, with a partner, or in a small group. Share your presentation and what you have learned with the class.

1. Research the Hubble Space Telescope and write a report about how the telescope works, and what some of the observations are that this telescope allows us to make.

2. Pick three everyday objects, such as coins, leaves, rocks, or even your skin, and use a hand lens to study them. Make sketches of each object. Label your illustrations to point out features that were visible with the hand lens.

3. Research tools that are used in the medical field to help scientists gather data about the human body. Write descriptions of three tools and explain how they work.

Chapter 3

Tools for Recording and Communicating

What tools are used to record and communicate information?

Scientists collect information using a variety of tools. But part of the scientific process is also communicating what they have learned. Scientists have a variety of tools to do that.

Models

Have you ever heard the expression “A picture is worth a thousand words”? If so, you probably know that it means that a picture can provide as much information as many written or spoken words. A scientific **model** is like a picture. It presents information in a visual form. A scientific model can be a picture, diagram, computer-generated image, or other representation of an object or process. Scientific models help scientists explain things that cannot be observed directly. For example, we can make models of things that are very large or very small.

Essential Vocabulary

bar graph	page 39
diagram	page 40
line graph	page 39
model	page 35
pie chart	page 39
table	page 38

A model of the solar system can describe the position of planets in space and their revolution around the sun. Such a model can be a drawing or a three-dimensional structure that looks like the real thing. This type of model is called a physical model.

Another example of a physical model is the model of an atom, the building block of matter. The atom is much too small to see, but scientific models can describe and explain its structure. This is also true of the models of plant and animal cells.

Sometimes a scientific model is not a drawing or three-dimensional structure, but rather a mathematical equation that describes how something works. Such a model is called a mathematical model. One example of a mathematical model is Einstein's equation for energy: $E = mc^2$. The chemical equation for the decomposition of water is another:

$$2H_2O \longrightarrow 2H_2 + O_2$$

▼ **The Franklin Institute in Philadelphia has a giant model of the heart that you can walk through.**

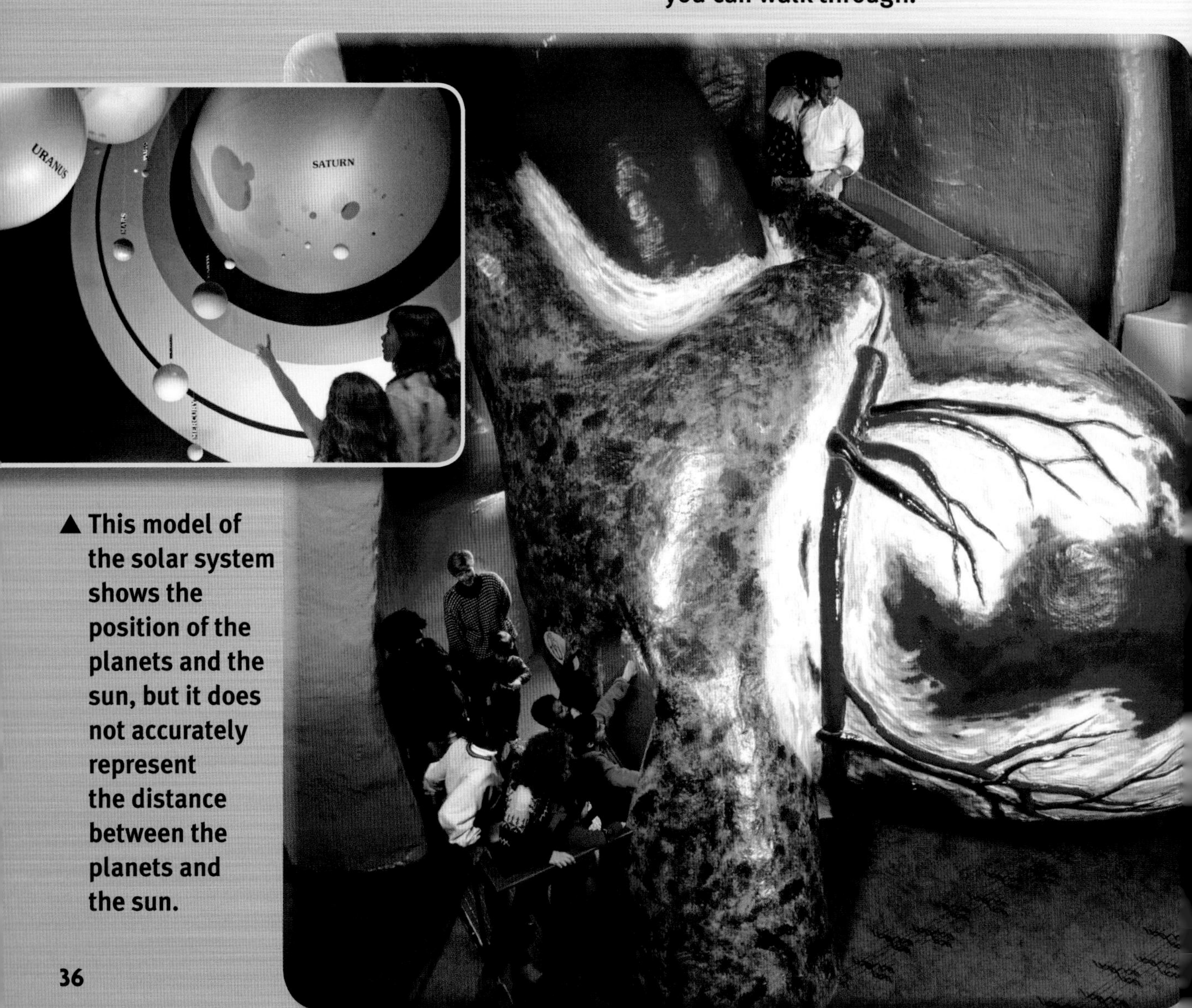

▲ **This model of the solar system shows the position of the planets and the sun, but it does not accurately represent the distance between the planets and the sun.**

Computers

Today, computers are part of every scientist's tool kit. They are used for many different jobs. Some are used to create reports. Others record data. Others do very complex tasks.

One way that scientists use computers is to create models. Computer models help scientists work with complex data. Computers use and analyze data and then can create images and models to study that data. Meteorologists use computer models to help them study hurricanes. It lets them predict the strength and path of a hurricane. Factors, such as water temperature, air temperature, and wind direction, affect the strength and path of a hurricane. Computer models let scientists see how each of these factors will affect the outcome. This tool helps them make predictions and warn people about hurricanes. What will happen if the water temperature rises by a degree? What will happen if the wind shifts to the east? What will happen if the wind shifts to the west?

Computer models are important tools in the study of climate change, one of the most challenging scientific issues today. There are many different factors involved in climate change. Scientists are gathering climate data from around the world. Computer models are helping them analyze this information.

Science and History

When ENIAC, or the Electronic Numerical Integrator and Computer, was unveiled in 1946, scientists were amazed by what the early computer could do. The computer, the size of a large room and weighing over 30 tons, could do mathematical calculations in seconds. The work would take a person hours or even days to do. ENIAC was developed for the army and it was used in a variety of projects—everything from wind tunnel design to weather prediction. Today, computers are much smaller and much faster. It was the invention of the microchip in 1959 that led to smaller and smaller computers.

Tables, Graphs, and Diagrams

Tables, graphs, flow charts, and diagrams are all tools used to communicate information.

Table

A **table** is one way to record and present information. A table on page 8 shows the different base units in the SI system. Scientists can also use tables to record information collected during an experiment. For example, a scientist that is tracking the growth of three different types of plants can use a table to record the results. Tables have columns that are vertical, and rows that are horizontal. A table can have as many columns and rows as needed. The table below shows the water temperature in three different locations on the Atlantic coast for six months of the year. More columns could be added to show all twelve months, and more rows could be added to list more locations.

Atlantic Ocean Water Temperature (°Celsius) January to June

	January	February	March	April	May	June
Bar Harbor, ME	3.3°	2.2°	3.3°	5.5°	8.3°	10.5°
Cape May, NJ	2.7°	2.7°	5.5°	8.8°	12.7°	19.4°
Miami Beach, FL	21.6°	22.7°	23.8°	25.5°	26.6°	28.8°

Pie Chart

A **pie chart**, also called a circle graph, shows the relationship of parts to a whole. The entire circle—the pie—represents all of the data. Each part of the circle—a slice of the pie—represents a category of the data. Look at the pie chart. It shows the percentages of different gases that make up Earth's atmosphere.

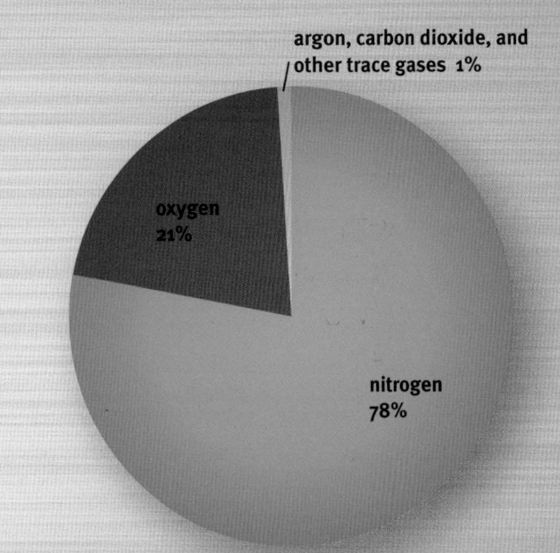

Line Graph

A **line graph** is most often used to show continuous change. The line graph shown here shows how the temperature changes at different levels of Earth's atmosphere. The y-axis, or vertical axis, represents the altitude, or distance from Earth's surface. The x-axis, or horizontal axis, represents the temperature.

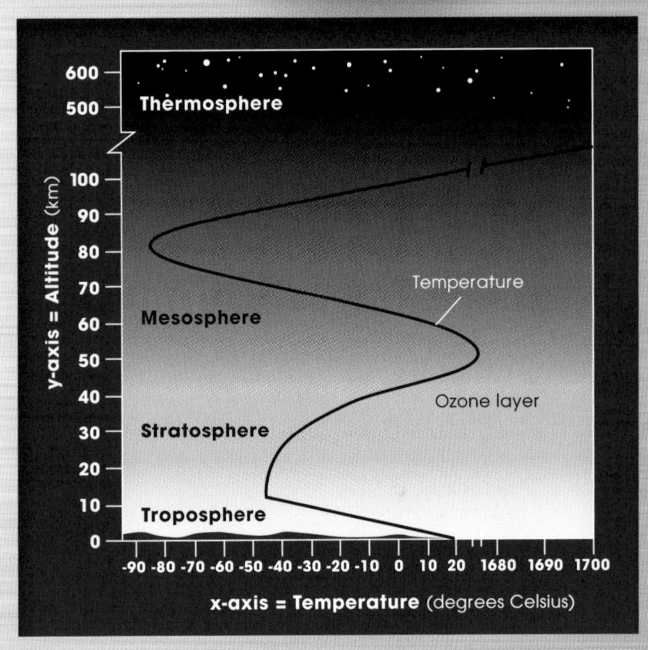

Bar Graph

A **bar graph** is another way to compare figures. It is used mainly to show change that is not continuous. This type of graph can be used to indicate trends for data that are taken over a long period of time. The bar graph shows the average monthly temperatures in Death Valley, California, over the period of a year. By showing the average for each time period, the graph allows a person to compare the information.

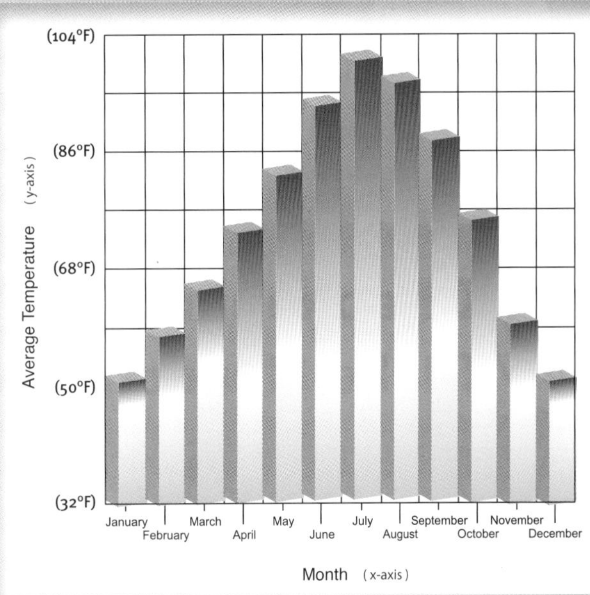

Diagrams

Diagrams are visuals that explain complex information or show relationships among things. A diagram of the water cycle illustrates how water is constantly moving among the oceans, atmosphere, land, and living things. A food web shows how energy is transferred in an ecosystem.

A diagram can also show how something works. For example, a diagram can illustrate the structure and operation of a telescope, camera, or microscope. You learned about the parts and operation of a compound microscope from the diagram on page 29.

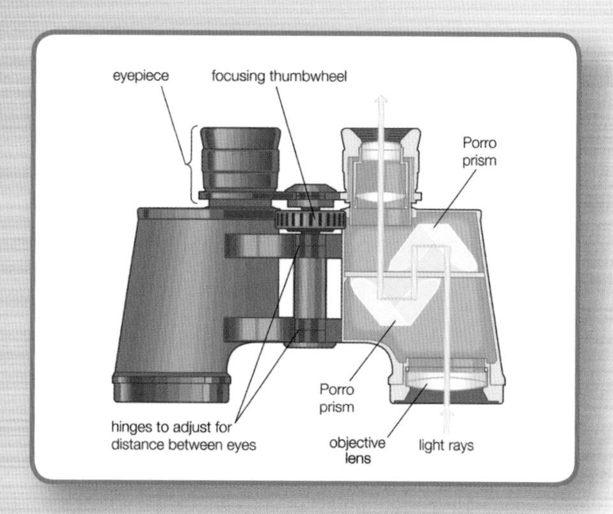

Some diagrams show the steps in a process. How a plant develops from a seed can be explained as a series of diagrams. So, too, can the delivery of electricity from a hydroelectric plant to a home.

The Water Cycle

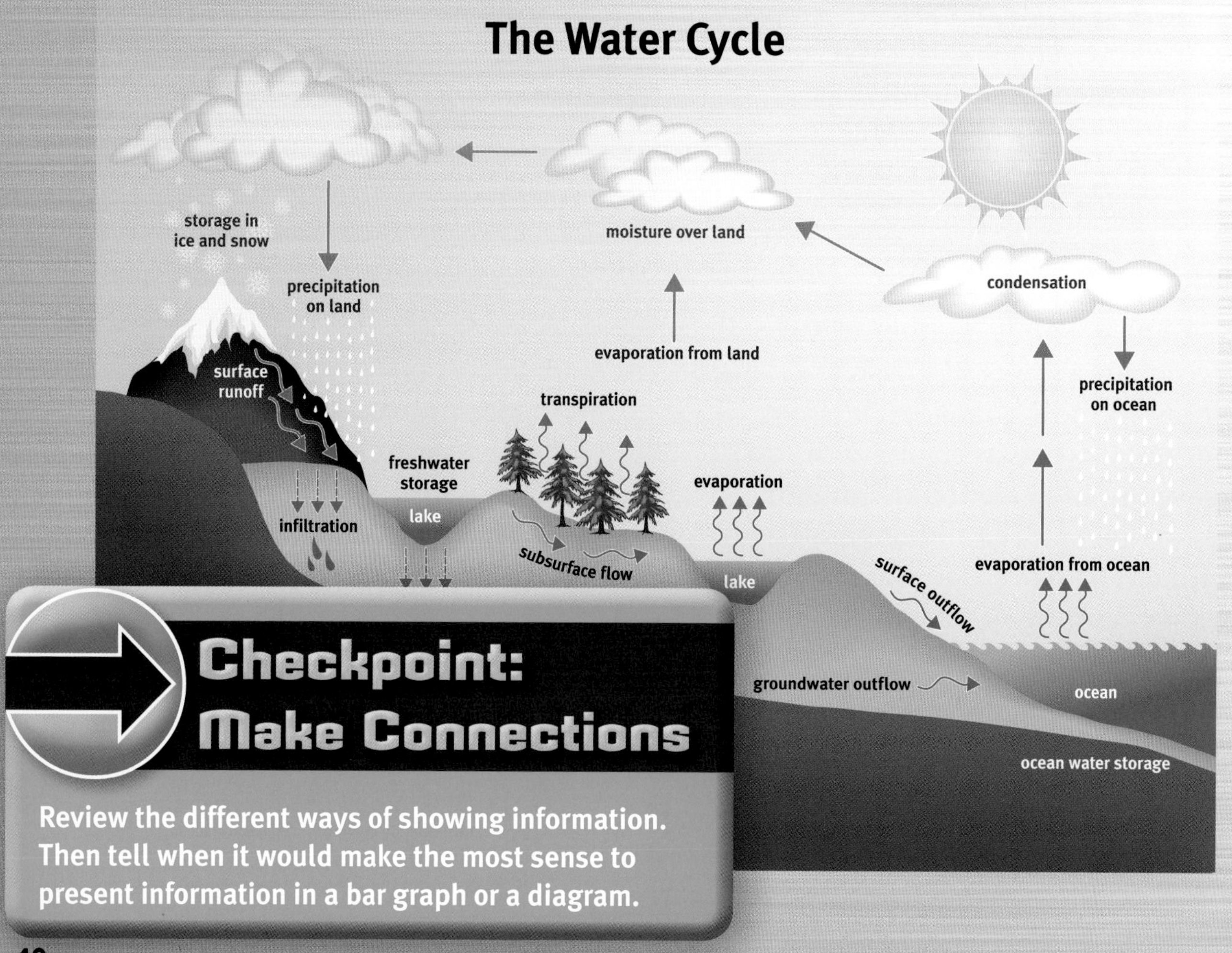

Checkpoint: Make Connections

Review the different ways of showing information. Then tell when it would make the most sense to present information in a bar graph or a diagram.

SUMMING UP

- Scientists use a variety of tools to help them record and communicate information.
- A scientific model is a picture, diagram, computer-generated image, or other representation of a complex object or process.
- Some models are physical models and some are mathematical models.
- Computer models can help scientists analyze data and work more easily with complex statistics.
- Tables, graphs, models, and diagrams are all visual ways for scientists to present information.

Putting It All Together

Choose one of the activities below. Work independently, with a partner, or in a small group. Share your presentation and what you have learned with the class.

1. Gather data about the weather in your community for one week. The data can include temperature, humidity, barometric pressure, wind speed, and wind direction. Prepare a table to summarize your information. Then use the table to construct an appropriate graph to communicate the information.

2. Collect samples of tables and graphs from newspapers and magazines. Make a poster display of your findings.

3. Research one of the computer models described in the chapter and then present your findings in an oral report.

Conclusion

Tools for a Scientific Expedition

The scientists on the expedition to the Celebes Sea made exciting discoveries. To do their job, they used many different tools.

Some of the tools the scientists used were for taking measurements. Satellites measured the surface water temperature in the Celebes Sea and in the surrounding water. Tools aboard the ROV measured the depth, temperature, and salt content of the water. Scientists used rulers to measure the length of specimens.

The scientists also used tools for observing. Scientists went scuba diving and used the Video Plankton Recorder to observe the animals in their natural habitat. The ROV cameras helped them observe marine life in the deepest water. The animals brought back to the research vessel were studied with a microscope.

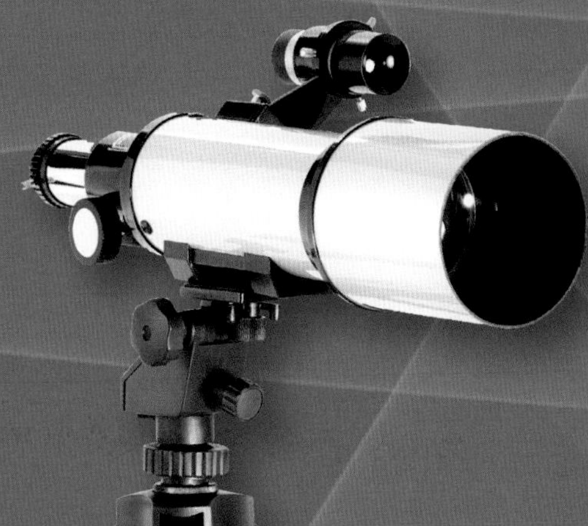

▲ Scientists used microscopes on the ship to study newly collected specimens.

Computers were essential tools for recording all of the data the scientists collected. At the end of the expedition, all of the information was analyzed. To share their observations and conclusions, the expedition scientists presented the information in tables, graphs, and diagrams.

The scientists were able to identify new species of deep-sea animals and find out more about this unusual area in the ocean. None of their work would have been possible without tools. Look back at the introduction. What other tools did the scientists use, and what jobs did the tools do? What other tools did the scientists use to measure, observe, and communicate information?

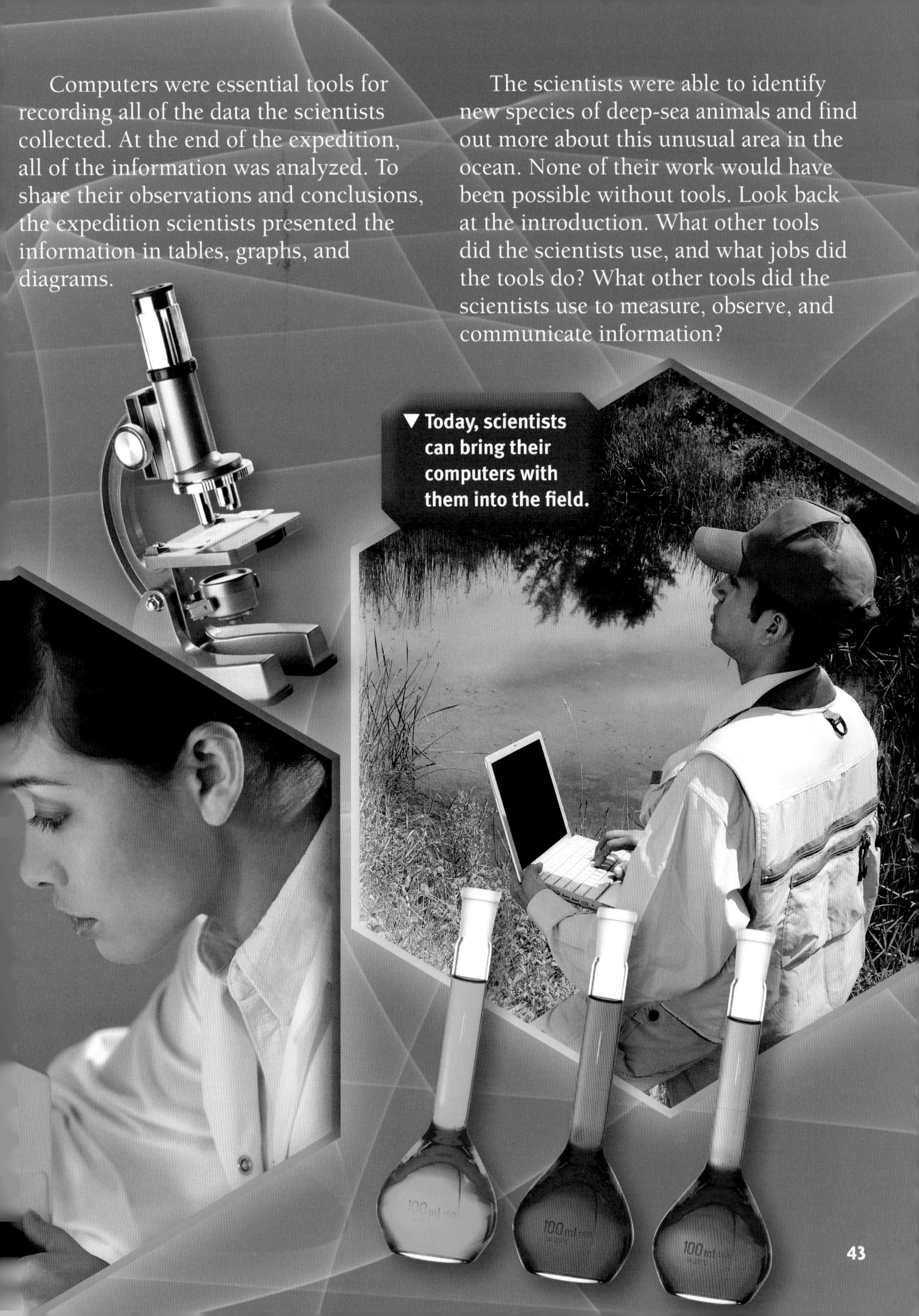

▼ **Today, scientists can bring their computers with them into the field.**

How to Write

When you are researching a scientific topic, you may find that the best way to organize and present the information you gather is in a table. A table is a good way to present data in an easy-to-read format. After you have completed a science laboratory investigation in which you've measured a variety of objects, a table is an effective means of recording the information. Tables are also helpful in comparing information.

To create a table, the first thing you must do is decide how to organize the information. Think about a useful way to sort the information into different categories. If you have researched minerals, you can make a simple table that lists the minerals in one column and describes them in another column. But if you have also gathered information about the density, hardness, and luster of each mineral, you will need to add a column for each of those qualities. Remember that each column in the table must have a title.

Sometimes a table has pictures, too. You can add a picture of each type of mineral. Finally, make sure you give your table a title.

The table below summarizes the three types of rocks. It has three columns and is organized to provide information about the formation of each type, and gives some examples. What other information could the writer include?

Three Types of Rocks

Type of Rock	Formation	Examples
sedimentary	forms over long periods of time as layers of sediment are pressed together	limestone, sandstone
igneous	forms from volcanic material that cools deep inside Earth or on Earth's surface	obsidian, pumice
metamorphic	forms from sedimentary rocks or igneous rocks that have melted and cooled again	marble, slate

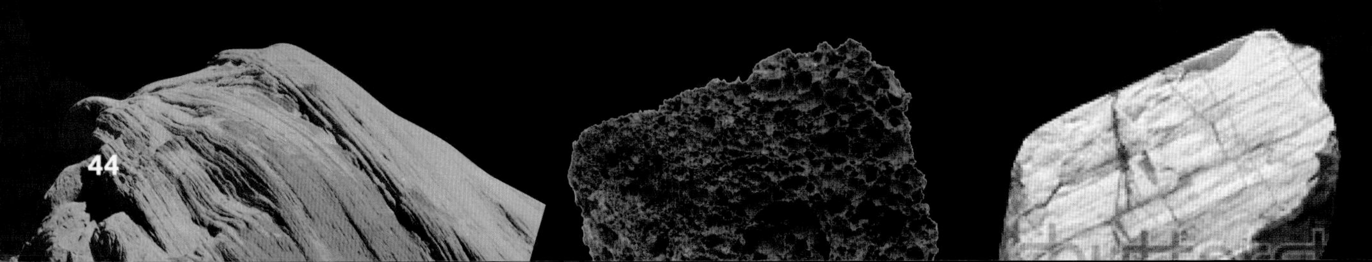

a Table

Look at the following notes from a report about vitamins and minerals. How could you turn this information into a table?

Vitamin A
- Needed for healthy skin, eyes, and immune system
- Found in liver, carrots, sweet potatoes, pumpkins, apricots, peaches, and papayas

Vitamin D
- Needed for healthy bones
- Helps body absorb calcium
- Body makes vitamin D when exposed to sun; also found in egg yolks and fish oils

Vitamin E
- Protects cells from damage
- Found in vegetable oils, nuts, and green, leafy vegetables

Vitamin B12
- Makes red blood cells
- Keeps nervous system healthy
- Found in meat, fish, and poultry

Vitamin B6
- Helps nervous system work properly
- Helps make red blood cells
- Found in potatoes, nuts, seeds, bananas, and spinach

Thiamine (B1)
- Needed for healthy brain and nervous system; helps body break down carbohydrates
- Found in wheat germ, sunflower seeds and other nuts, and oatmeal

Niacin (B3)
- Helps body break down carbohydrates, protein, and fat
- Found in red meat, poultry, fish, and peanuts

Riboflavin (B2)
- Helps body break down carbohydrates, protein, and fat
- Needed for healthy eyes, skin, hair, and nails
- Found in milk, eggs, and almonds

Folic Acid (Folate/B9)
- With B12, helps protect nervous system
- Helps with production of DNA
- Found in dried beans, lentils, chickpeas, and dark green, leafy vegetables

Calcium
- Needed to build strong bones and teeth
- Found in milk and other dairy products such as yogurt and cheese, and dark, green leafy vegetables such as broccoli

Iron
- Helps red blood cells carry oxygen
- Found in red meat, lentils, and spinach

Magnesium
- Helps muscles work well
- Needed for energy
- Found in whole grain breads, nuts, and seeds

Potassium
- Helps body maintain right balance of water
- Helps muscles and nerves function
- Found in broccoli, potatoes, citrus fruits, and bananas

Zinc
- Helps immune system
- Promotes healing
- Found in red meat, poultry, seafood, nuts, milk, and dairy products

bar graph (BAR GRAF) *noun* a diagram that represents different values by bars of different heights (page 39)

binoculars (bih-NAH-kyuh-lerz) *noun* a tool that makes faraway objects appear close (page 26)

Celsius (SEL-see-us) *adjective* a unit of measure for temperature; 0 is the freezing point and 100 is the boiling point on the Celsius scale (page 18)

cubic centimeter (KYOO-bik SEN-tih-mee-ter) *noun* a derived unit used to measure volume (page 14)

density (DEN-sih-tee) *noun* the amount of mass in a given volume (page 17)

diagram (DY-uh-gram) *noun* a visual representation of a process or an object (page 40)

hand lens (HAND LENZ) *noun* a small, handheld tool used to magnify objects (page 27)

kelvin (KEL-vin) *noun* the base unit of measurement in the SI system for temperature; the Kelvin scale is based on absolute zero (page 18)

kilogram (KIH-luh-gram) *noun* the base unit of mass in the SI system of measurement (page 10)

line graph (LINE GRAF) *noun* a diagram that represents the relationship between two variables (page 39)

liter (LEE-ter) *noun* a unit of measure for volume; accepted for use in the SI system (page 13)

mass (MAS) *noun* the amount of matter an object has (page 10)

meniscus (meh-NIS-kus) *noun* the curved surface of a liquid in a graduated cylinder, or other container (page 13)

meter (MEE-ter) *noun* the base unit of length in the SI system (page 9)

metric system (MEH-trik SIS-tem) *noun* a system of measurement that uses units of ten; also called the SI system (page 8)

microscope (MY-kruh-skope) *noun* a tool that produces a magnified image of an object; used to view objects that are too small to be seen with the naked eye (page 27)

model (MAH-dul) *noun* a physical or conceptual representation of an object, a theory, or a process (page 35)

newton (NOO-tun) *noun* a derived unit of weight in the SI system; used to measure force (page 12)

pie chart (PY CHART) *noun* a circular graph divided into parts that represent proportions of the whole (page 39)

reflecting telescope (rih-FLEK-ting TEH-leh-skope) *noun* a telescope that uses mirrors to gather and focus light (page 24)

refracting telescope (rih-FRAK-ting TEH-leh-skope) *noun* a telescope that uses lenses to gather and focus light (page 24)

sonar (SOH-nar) *noun* a tool that sends sound waves through water or air; used to measure the depth of water and to find objects under water (page 31)

table (TAY-bul) *noun* a data or informational chart usually arranged in rows and columns for easy reference (page 38)

telescope (TEH-leh-skope) *noun* a tool that makes very distant objects appear closer by; used for gathering and focusing light (page 24)

volume (VAHL-yoom) *noun* the amount of space an object takes up (page 13)

Index